PROJECT SCIENCE

Energy and Forces

Terry Jennings
Illustrated by **David Anstey**

CONTENTS

SMITHMARK

ON THE MOVE

All around us things are moving. Nothing can move unless a *force* starts it. Most forces are pushes or pulls. Twists, tugs, presses, squeezes, stretches, and twists are also forces. They all move things.

Look at these pictures. What is making each of the things in the pictures move? Is it a pull, or is it a push? Is it a stretch, or a squeeze? Or is it more than one of these things?

PUSH OR PULL?

Look around your home or school. Think carefully about the things you move about. Do you move them with pushes or pulls, or both?

Make a chart like this. Fill in how each of the things is moved. Add as many things to the chart as you can.

Pushes	Pulls	Pushes and pulls
Light switch	Curtains	

TUG OF WAR

1 Organize two teams of children. Make sure they have equal numbers and sizes. On the word GO both teams should pull hard on their ends of the rope. What happens if both teams' pulls are equal?

2 Ask one heavy person at one end of the rope to stop pulling. What happens now that you have unequal forces?

Backward and Forward

Forces always act in pairs. When you push a sledge, it pushes back at you with an equal but opposite force. If your feet are not firmly on the ground you may slip over.

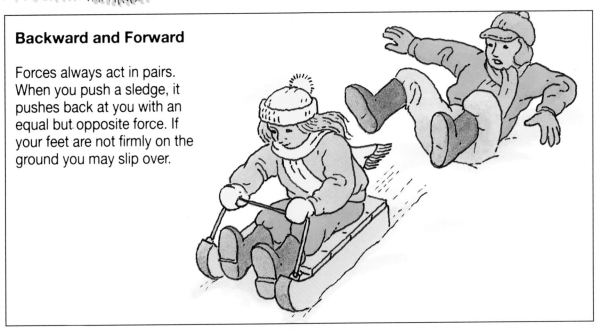

5

CHANGING DIRECTION

All machines work because of pushes and pulls. Some car engines pull the cars forward. Others push the cars forward. It is not always easy to tell just by looking.

Ships are pushed along by propellers at the back. Look at these two pictures. Are these machines pushing or pulling?

ALL CHANGE

Most pushes move something away from you. Most pulls bring something toward you. But does this always happen?

1 Push a toy truck. Which way does it go? Does it move toward you or away from you?

2 Tie a piece of string to the truck. Pull on the string. Which way does the truck go? Does it move toward you or away from you?

3 Now loop the string around a table leg. Pull on the string. Does the truck move toward you or away from you?

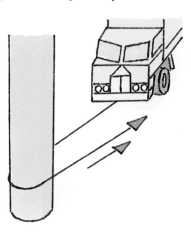

TAKING THE STRAIN

Pulleys change the direction of a pull. You can make your own simple pulley.

You will need: 2 pieces of thick wire about 8 in (20 cm) long; 2 empty spools; toy bucket; toy bricks; strong string.

1 Ask an adult to help you to bend the wire into a triangle shape. Push the ends into the holes in the spool.

2 Hang your pulley from a nail or a hook in a shelf.

3 Tie one end of a piece of strong string to the handle of a toy bucket. Fill the bucket with toy bricks.

4 Lift the bucket straight up by pulling on the string. How easy is it to lift the bucket?

5 Now loop the string over your pulley. Lift the bucket again by pulling on the string. Is it easier or more difficult to lift than before?

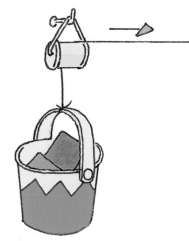

6 Make a double pulley using some more wire and another spool. Thread the string around the two pulleys.

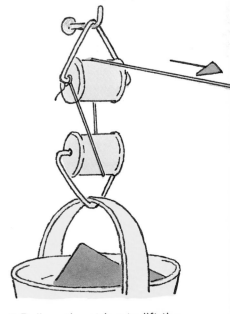

7 Pull on the string to lift the bucket. Is it easier or more difficult to lift than it was with the single pulley?

Lighten the Load

Pulleys make lifting easier. Cranes use pulleys to help them lift heavy loads. Can you see the pulleys on the crane in the picture? A motor provides the power to pull the cable over the pulleys and so lift the load.

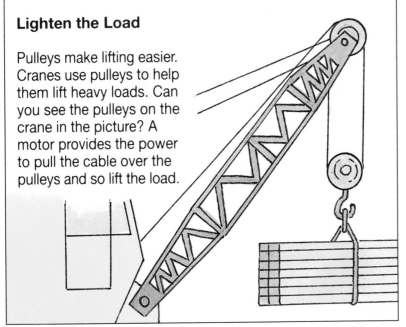

SLIDING AND ROLLING

One way to move things is by sliding them along. We often slide heavy boxes and boats because they are too heavy to lift.

Many moving things have wheels which help them to roll along. Cars, bicycles, and shopping carts all have wheels so that they roll along more easily.

RUBBER POWER

You will need: a piece of wood; cup hook; large rubber band; brick; pen; pencils.

1 Fix the hook in the center of one side of a piece of wood. Mark two dots on a rubber band a few inches apart. Loop the band over the hook.

2 Put a brick on the piece of wood. Pull it along using the rubber band. Measure the space between the dots. How much does the rubber band stretch before the brick and wood start moving?

3 Now lay the brick and piece of wood on the pencils. Pull the brick and wood along with the rubber band. How much does the rubber band stretch? Is it more or less than before?

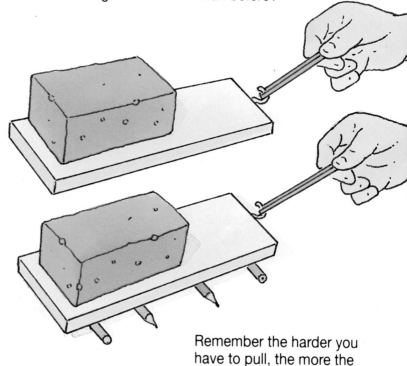

Remember the harder you have to pull, the more the rubber band stretches. Which is easier, sliding or rolling?

A SLIPPERY SLOPE

Find out which is easier, sliding or rolling.

1 Fill a tin can or coffee can with sand. Put the lid on the can. Raise one end of a plank of wood on books to make a ramp.

2 Stand the tin can at the top of the ramp. Give it a gentle push. How far does it slide?

3 Now lay the tin can at the top of the ramp. You might need to give it a gentle push. How far does it move? Is it farther than before?

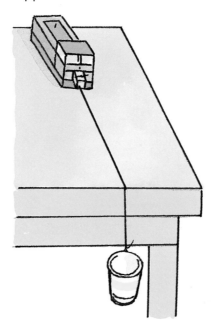

WHEELING ALONG

You will need: cardboard shoe box or plastic building blocks; a piece of string; small plastic cup; toy bricks; weights or marbles.

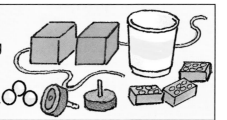

4 Put the weights, or marbles, in the plastic cup, one at a time. How many weights do you have to add to make the truck slide from one end of the table to the other?

1 Make a toy truck using the building blocks or a cardboard shoe box but do not put the wheels on yet.

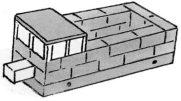

2 Tie one end of a piece of string to the truck. Tie a small plastic cup to the other end. Put a few toy bricks in the truck to make it heavier.

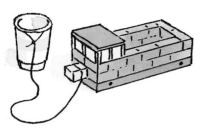

3 Stand the truck at one end of the table. Hang the string and plastic cup over the opposite end of the table.

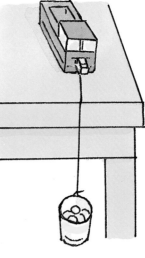

5 Now start again, but this time put the wheels on the truck. How many weights do you need to make the truck roll from one end of the table to the other? Is it more or less than before? Which is easier, sliding or rolling?

SLIDING

When two surfaces rub together, a force called *friction* stops them from moving easily against one another. It makes moving much more difficult.

Sometimes sliding is easier than rolling. For example, car tires cannot grip well in deep snow or on icy roads. But a sled slides easily over slippery ice or snow. On the other hand, it would be much more difficult to pull a sled along a dry road.

EASY MOVERS

Make a collection of objects that are all roughly the same size. A stone, a piece of wood, an ice cube, an eraser, and a piece of sponge are some of the things you could try.

1 Use a plank of wood and some books to make a ramp. Put the stone at the top.

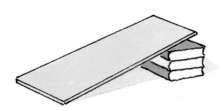

Give the stone a gentle push. Watch how far and how fast the stone travels. Measure the distance.

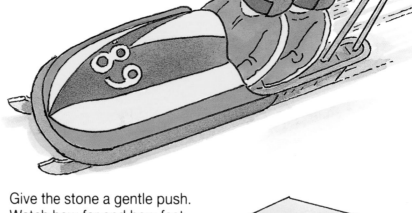

2 Now try each of the other objects. Which of them move most easily? Are they the objects that cause less friction?

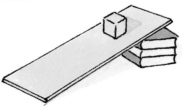

3 Feel all of the objects. What do you notice about those that have less friction?

Repeat the experiment, but this time wet the plank first. What differences do you notice?

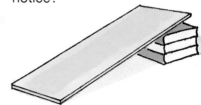

BRAKES ON

What effect does friction have on rolling objects?

You will need: a plank of wood; toy car; sheets of newspaper; books; rug or a piece of carpet; some oil.

1 Make a ramp on a smooth floor.

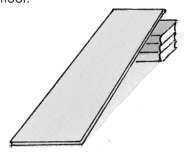

2 Put a toy car at the top of the ramp and let it go. Measure how far the car travels.

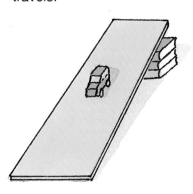

3 Now put some sheets of newspaper at the bottom of the ramp. Let the car roll again. How far does it go? Is it more or less than before?

4 Try the experiment with a rug or piece of carpet at the bottom of the ramp. Does this make a difference? Make a chart of your results.

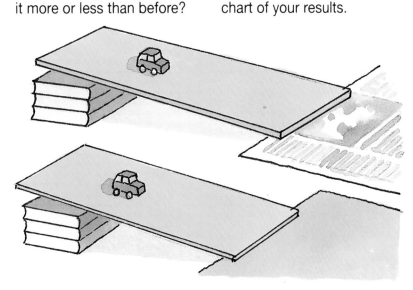

nothing
newspaper
carpet

5 Put a tiny drop of oil on the axle of each wheel. Repeat the experiment. What difference does the oil make?

Look at your chart. When is there most friction?

Gripping Stuff

Sometimes friction is useful. It helps us to grip when we walk or run. It also helps the wheels of cars and bicycles to grip the ground.

Sometimes friction is a nuisance. The rubbing wears out your shoes, clothes, tires, and some machinery. People put oil on their machines to try to reduce friction and stop parts from rubbing and wearing out.

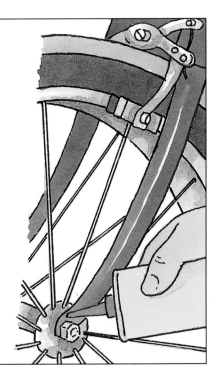

STOP/GO

It takes more force to start a heavy object moving than to make a light object move. It also takes more force to stop heavier objects. This is because heavy objects have more *inertia*, which resists starting and stopping.

STOP

What happens to moving objects when they stop suddenly? You can find out in this simple experiment.

You will need: a roller skate or a toy truck; small doll; cardboard box.

1 Put a small doll on the skate or in the truck. Push the skate or truck to make it move. What happens to the doll when the truck starts moving?

2 When does the truck stop moving? What happens to the doll when the truck stops?

3 Put the doll back in the truck. Push the truck so that it bumps into a cardboard box. What happens to the truck? What happens to the doll?

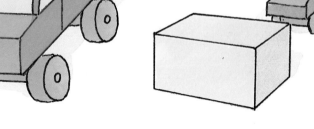

Safety First

The children in the picture are sitting in the back seat of a car. They are not wearing seat belts. Think about what you have discovered. What will happen to the children if the car moves off suddenly? What will happen if, when the car is moving, it stops suddenly?

How can you keep the doll from moving when the truck starts or stops moving?

READY, SET, GO!

Is it easier to start something moving quickly or slowly?

You will need: a toy truck; 2 heavy books; length of cotton thread.

1 Rest two heavy books on the truck. Tie a length of cotton thread to the front of it.

2 Pull gently but steadily on the thread. Does the loaded truck start moving?

3 Let the thread go slack. Now give it a really hard tug. What happens to the thread?
  If you were fishing and caught a really big fish, would you pull it in slowly or quickly? Why?

STAYING POWER

1 Fix a large rubber band to the front of a toy truck. Make two ink spots a few inches apart along one side of the rubber band.

2 Pull the rubber band steadily until the truck starts moving. All the time look carefully at the ink spots on the rubber band. What happens to them if you pull harder?

When do you have to pull hardest?

13

G-FORCE

There is another force that we meet every day. Have you ever wondered why, when something falls, it always falls down to the ground? It does so because the earth has an invisible force that pulls other objects towards it. This force is called *gravity*.

To make something start or stop moving, you must overcome its inertia – usually by pushing or pulling it. The heavier something is, the harder you have to push or pull to start it or stop it from moving. But gravity pulls things at the same speed, no matter what they weigh. To test this for yourself, try the projects below.

FALLING DOWN

1 Lay a metal tray on the floor in front of a chair. Stand on the chair. Hold a small, heavy ball bearing in one hand and a round bead in the other. Hold both hands high above your head.

2 Drop the ball bearing and the bead at the same time. Which hits the tray first? Is it the heavy ball bearing or the lighter bead?

3 Try other pairs of objects. Choose ones that are the same shape but different weights. You could try a tennis ball and a golf ball. Which hits the ground first? Try a large and a small coin. Which hits the ground first?

Newton's Apple

Gravity was first explained by a famous scientist called Isaac Newton in the 17th century. Newton was puzzled as to why, when an apple falls from a tree, it falls down to the ground. He decided that the earth has a powerful force that is able to pull the apple down to the ground. He called this force gravity.

DEAD WEIGHTS

Things have *weight* because gravity pulls on them. The more gravity pulls on an object, the more it weighs.

You can make a simple weighing machine or balance from a thick rubber band.

You will need: a thick rubber band; thumb tack; paperclip; yogurt cup; thin string; ruler; marble; eraser; small pebble.

ALWAYS ask permission to use household objects.

1 Push a thumb tack into the edge of a shelf.

2 Loop a paperclip on the rubber band. Hang the paperclip from the thumb tack.

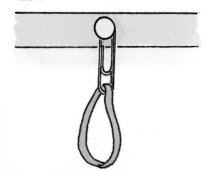

3 Make three small holes around the edge of a yogurt cup. Thread three lengths of thin string through the holes and tie them to make a handle. Fix the handle to the bottom of the rubber band.

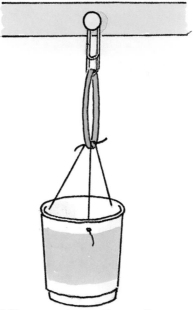

4 Tape a ruler to the wall behind the rubber band or draw a scale on a strip of poster board and tape this to the wall.

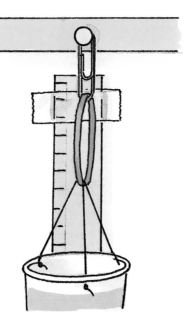

5 Bend a paperclip to make a pointer and fix it to the rubber band.

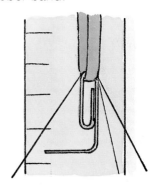

6 Make a note, or mark, where the pointer is on the scale when the cup is empty.

7 Then put a small object such as a marble, an eraser, or a small pebble in the yogurt cup. Mark how far the rubber band is stretched down by the weight.

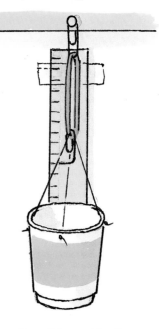

Try other things. Do they weigh more or less?

If you have some real weights you could use these to see that your balance weighs accurately.

FUN FORCES

We have looked at gravity, friction, and inertia. There are lots of everyday forces. They include stretching, twisting, spinning, and swinging. We can use these forces to work some toys you can make for yourself.

QUICK RETURN!

Make this moving tin. Show it to your friends. Tell them that it is magic.

You will need: a large tin or coffee can with a lid; hammer; nail; large rubber band or narrow elastic; a weight such as an iron nut; cotton thread.

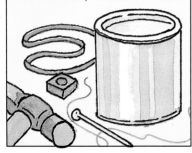

1 Ask an adult to make two holes in the lid and two more in the bottom of the tin, using the hammer and nail.

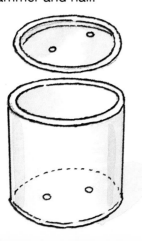

2 Cut a large rubber band in two so that it makes one long piece of rubber. Or cut a piece of narrow elastic.

3 Thread the piece of rubber or elastic through the holes in the tin. See that it crosses over in the middle. Tie the ends together.

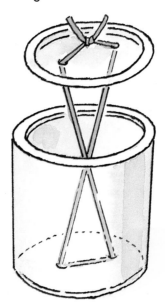

4 Ask someone to hold the lid of the tin up, stretching the rubber or elastic, while you tie a weight inside the can.

5 Push the lid on the tin. Roll it gently away from you across the floor. Watch what happens. Can you explain what you see?

BACK AND FORTH

Swinging is another type of movement started by a push or a pull. You can discover more about swinging if you make a simple *pendulum*.

You will need: a thick piece of wood; thumb tack or hook; empty spool; piece of string; clock or watch with second hand; 2 iron nuts or some other heavier weight.

1 Push a thumb tack or a hook in the side of the piece of wood and lay it on an old table.

2 Tie a spool onto a piece of string. Loop the other end around the hook or thumb tack.

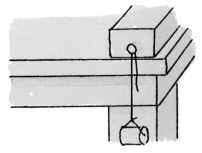

3 Borrow a clock or a watch with a second hand. Set your pendulum swinging. How many swings does it make in 10 seconds?

4 Make the string longer. Now see how many swings the pendulum makes in 10 seconds. Is it more or less than before? Or is it the same?

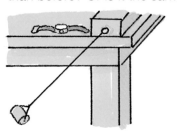

5 Start again with a heavier weight on your pendulum. What difference does this make?

Look at the pendulum in a clock. What is it made of? What does it do?

Way Out

When something moves in a circle, it tries to move outward. You can test this for yourself.

1 Put a tennis ball in a small net bag. The kind oranges are sold in will do well.

2 Tie a length of string to the bag.

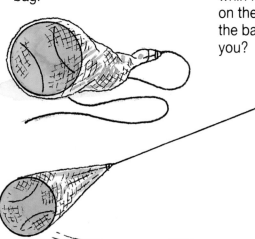

3 Take the ball outside and whirl it around and around on the string. Can you feel the ball pulling away from you?

MUSCLE POWER

Everyday we use our bodies to make lots of forces. We push and pull. We twist and turn, we lift and carry. We do all these things with the help of muscles. Our bodies have hundreds of muscles. Nearly all of them work in pairs.

A MODEL ARM

You will need: a piece of poster board; paper fastener; thin string.

1 Cut out two arm shapes from poster board.

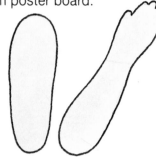

2 Join the two parts of the arm together with a paper fastener.

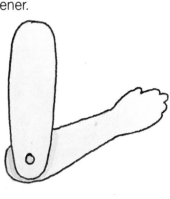

3 Attach two pieces of thin string to act as muscles as shown.

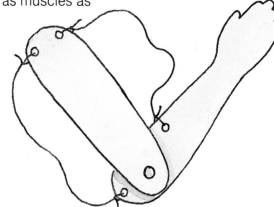

4 Pull on one of the pieces of string (string 1). What happens to the lower arm?

5 Now pull on the other piece of string (string 2). What happens to the lower arm?

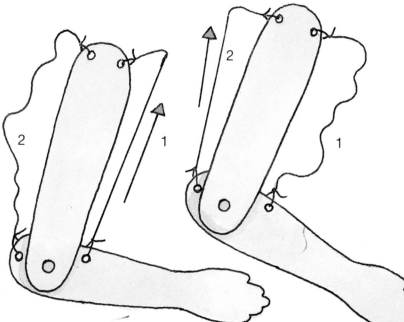

ARM MUSCLES

1 Find a heavy book. Slowly lift it up with one hand. Hold your other hand against the front of your upper arm.

Can you feel the muscle changing? It goes from being long and thin to being short and fat. We say the muscle has *contracted*. The muscle pulls up your forearm and the book in your hand.

2 Now put your hand on the muscle behind your upper arm. Slowly lower the book. What happens to that muscle? Can you feel it contracting? What happens to the muscle at the front of your upper arm? Can you feel it getting longer and thinner? We say it has *relaxed*.

Most muscles work in pairs like this. This is because muscles can only pull on bones. They cannot push on them.

PUSHING POWER

How hard can you push with one foot or one finger? Here is a way to find out.

1 Stand a set of bathroom scales against a wall.

2 Lie on your back on the floor. Push the bottom of one foot against the scales as hard as you can. Ask a friend to read the scales while you are pushing on them.

Now try with the other foot. With which foot can you push hardest? Try to find out which of your leg muscles are helping you to push.

3 Put the scales on a table. Push down on them with one finger. How hard can you push? Now try each of your other fingers in turn. With which finger can you push hardest? Make a chart of all your readings. Which reading is highest?

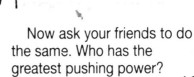

thumb
little finger
middle finger

Now ask your friends to do the same. Who has the greatest pushing power?

19

MOVING THINGS

A push or a pull is needed to move something. This takes *energy*. Moving things – people, animals, and machines – can do work because they have energy.

There are lots of different kinds of energy. Sound, light, electricity, heat, and movement are all forms of energy. You cannot see most of them. But you can see what they do in making things move.

You cannot use up energy. You simply change it from one kind to another.

Look at these pairs of pictures. The hammer, the bird, and the ball have energy. How do you know?

Suppose these things were not moving, as shown in the second picture of each pair. Then would they have as much energy?

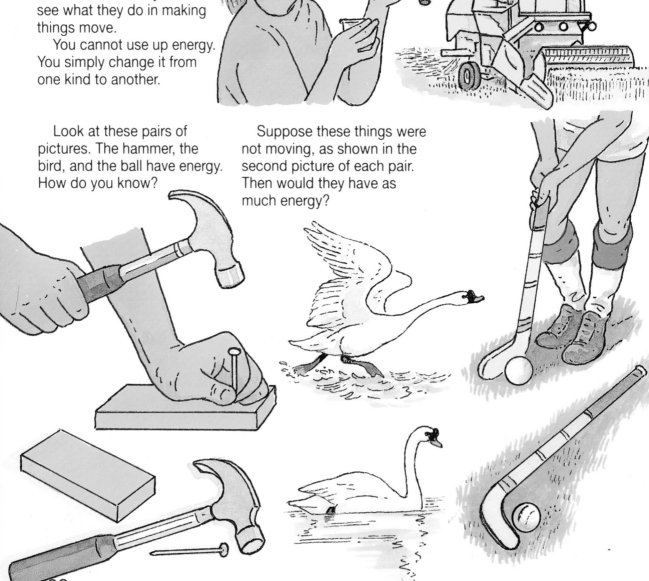

A WIND WHEEL

Make this simple wind wheel.

1 Make a hole in the center of the plate with the thumb tack.

2 Color the plate. Paint or crayon the four cups different colors.

3 Tape all four paper cups around the plate.

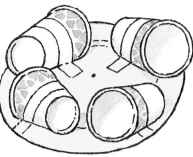

4 Use the thumb tack to fix the plate to the top of a short stick. Make sure that the plate turns easily.

5 Blow in one of the cups. Why does the wheel move? Where does the energy come from?

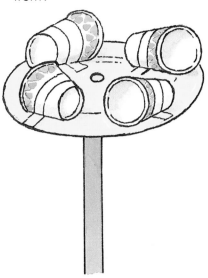

Take the wind wheel outside on a windy day. Does the wheel move? How many times does it turn in a minute? Where does the energy come from?

Turn Around

Look at these pictures. What makes these wheels turn? Can you think of another form of energy that could make a wheel turn?

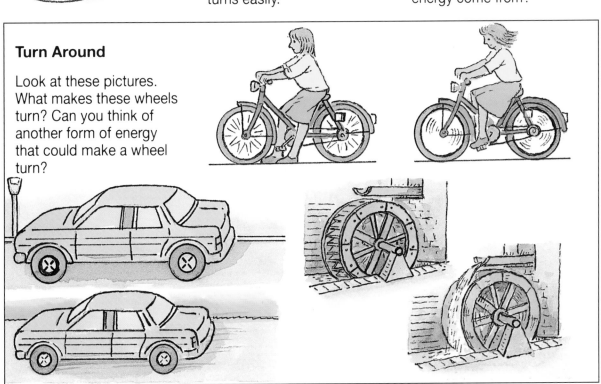

ON THE MOVE

Moving things have a kind of energy called *kinetic* energy. When we lift something up, it gains energy. This energy changes into kinetic energy when we let the object fall and it begins to move on its own.

Real windmills, including those used to produce electricity, are turned by the kinetic energy of the wind. There are many other ways of using energy to move things.

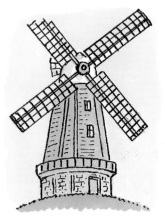

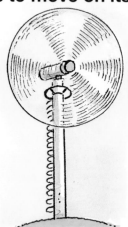

ROLLING ALONG

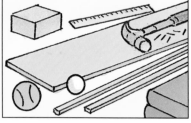

You will need: a plank; 2 thin strips of wood; books; golf ball or tennis ball; ruler; small cardboard box; small nails; hammer; ruler.

1 Nail a thin strip of wood along each of the long sides of the plank.

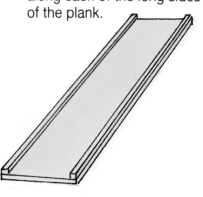

2 Rest one end of the plank on books to make a ramp, and put a small cardboard box at the bottom.

3 Hold a golf ball or tennis ball 12 in (30 cm) from the bottom of the ramp.

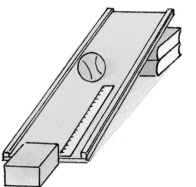

4 Let go of the ball. Watch it roll down and hit the box. Measure how far the box is pushed by the ball. Does the rolling ball have energy?

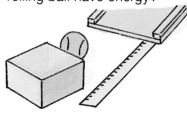

5 Put the box back at the bottom of the ramp. This time, roll the ball from the top.

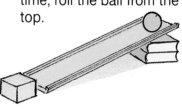

Now how far is the box pushed by the ball? Is it more or less than before? Did the ball have more or less energy than before? Where did the energy come from?

CLAY BALLS

1 Make two balls of the same size from modeling clay.

2 Stand on a hard floor. Drop one of the balls from shoulder height. Look at the ball carefully. How has it changed?

3 Now stand on a chair. Drop the other ball. Look at it carefully. How is it different from the first ball? Which ball had the most energy?

ROLLER COASTER

You will need: a long strip of poster board (about 3 in (7.5 cm) wide); thick string; glue or tape; ball bearing or marble.

1 Make a track from the long strip of poster board. Glue or tape thick string along both edges of it.

2 Fix one end of your track to a chair. Arrange it so that it has lots of hills and valleys in it.

3 Put a ball bearing or marble at the top of the track. Let it roll down the slope. How far does it go?

4 How can you make your ball bearing go right to the end of the track? Where does its energy come from?

Make another track and have roller coaster races with your friends.

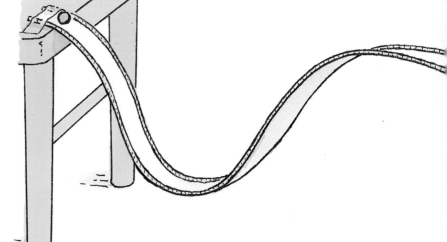

Up and Down

Have you ever seen the roller coaster rides at a fair? Where do the cars get their energy?

ENERGY STORES

Many objects and all living things (see page 32) have energy waiting to be used. This stored energy is called *potential energy*.

In the case of a ball held high and then dropped, the potential energy is changed to movement. The ball gets its energy originally because we use energy, from our muscles, to lift it up high.

Energy is never lost. It is simply changed from one kind to another. The higher you lift a ball or some other weight, the more potential energy it has. As it falls, the ball gets closer and closer to the ground and has less and less potential energy.

It would hurt much less if a weight fell on your toe from 12 in (30 cm) than from 90 feet (30 meters).

FLYING HIGH

You will need: a large rubber band; wire coat hanger; pliers or wire-cutters; plastic bottle; piece of candle; scissors; poster board; nail or paperclip; modeling clay.

Put the piece of candle, without its wick, in the neck of the bottle.

6 Cut out a small poster board airplane. Tie it to the end of the wire.

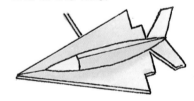

7 Wind up the rubber band using the wire. Let your plane go. Does it fly? You may need to balance it with little pieces of modeling clay or paperclips to get it to fly level.

1 Ask an adult to cut a piece off the coat hanger and bend one end of it to make a hook.

4 Fasten one end of the rubber band with a nail or large paper clip to the underside of the bottle. Use the hooked wire to pull the rubber band through the bottle.

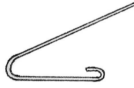

2 Cut a second, straight piece of wire and make a small hook in it.

3 Ask an adult to make a hole in the bottom of the bottle.

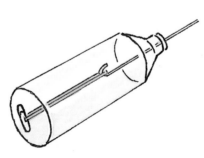

5 Put the bent wire on the other end of the rubber band.

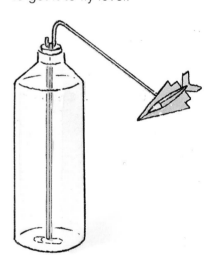

FROM TANK TO MERRY-GO-ROUND

You will need: a thread spool; paperclip; short piece of candle; cotton thread; nail; lollipop stick; rubber band; toy brick or block of wood; modeling clay; thin poster board; colored pencils.

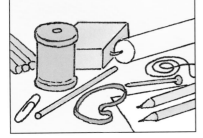

1 Cut a slice (about 1/2 in (1 cm) thick) off the candle. Pull out the wick and make the hole bigger with a sharp pencil or a nail.

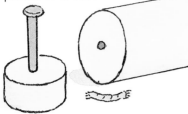

2 Push the rubber band through the holes in the spool and piece of candle.

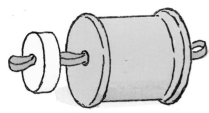

3 Fix the paperclip on the end of the rubber band next to the piece of candle.

4 Push part of the lollipop stick through the other loop in the rubber band. Make sure that the rubber band is not too tight so that the spool can move against the stick.

5 Wind up the toy tank by turning the lollipop stick. Turn it a few times and then put the tank down on the floor or table. Measure how far it goes. Can you make it go farther by turning the stick more times than before?

6 Make another model tank. But this time put the piece of candle at the same end as the lollipop stick.

7 Use modeling clay to attach your model to a toy brick or small block of wood.

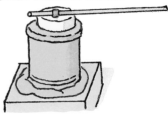

8 Cut a small horse out of poster board. Color it with your pencils and tie it to the end of the lollipop stick.

9 Wind the rubber band up using the lollipop stick.

10 Then let your model merry-go-round go. Does it work? Where is the energy coming from?

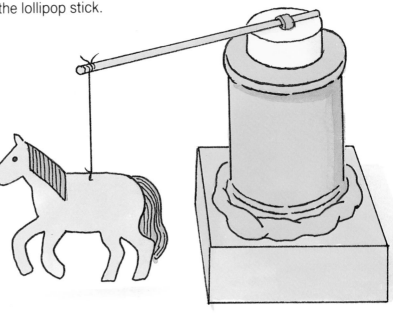

BEAUTIFUL BALLOONS

Jet engines and propellers make vehicles like aircraft and ships move by pushing air or water backward. The action causes the vehicle to move in the opposite direction, driving it forward.

BALLOON ROCKETS

Real rockets burn chemical fuels. They turn these fuels into gases which rush out of the ends of the rockets, pushing them high in the sky. Stored chemical energy is turned into movement energy.

Make your own balloon rocket and see how gases (air) rushing out of the balloon push it along.

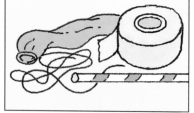

You will need: a balloon; strong thread or fishing line; tape; drinking straw.

1 Cut a drinking straw in half and put the two pieces on the thread or fishing line.

2 Tie the thread tightly between two hooks or the backs of two chairs.

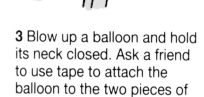

3 Blow up a balloon and hold its neck closed. Ask a friend to use tape to attach the balloon to the two pieces of drinking straw.

Let go of the balloon. Does it fly along?

4 Fill the balloon half-full of air. Measure how far it travels. Now blow the balloon up hard. How far does it travel?

Make two balloon rockets and have races with a friend.

A MODEL HOVERCRAFT

A hovercraft flies on a layer of air. This is blown down underneath the hovercraft by a powerful fan. Propellers steer the hovercraft. You can make a model hovercraft which is powered by a balloon.

You will need: a piece of poster board; nail; balloon; scissors; small cork; glue.

1 Cut a circle of poster board about 4 in (10 cm) in diameter. Make a hole in the center of the poster board with the nail.

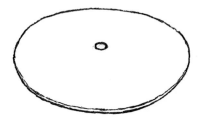

2 Find a small cork that will fit tightly into the neck of the balloon. Ask an adult to make a hole through the center of the cork with the nail.

3 Glue the cork to the circle so that the two holes are in line.

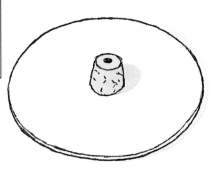

4 Ask a friend to hold the circle while you blow up the balloon. Keep the neck of the balloon pinched with the fingers of one hand while you stretch the mouth of the balloon over the cork.

5 Put your model on a smooth surface. Let go of the neck of the balloon. Give your hovercraft a gentle push.

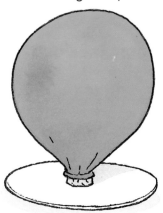

See how far your hovercraft will go if the balloon is only partly blown up and when it is blown up hard.

Will your hovercraft move over a rough surface? Will it fly over water? Make another hovercraft and have races with a friend.

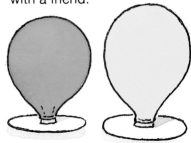

A Useful Craft

The hovercraft can travel over 6 feet (2m) above the surface. It is able to ride above seas, marshes, and even rough land.

HEAT POWER

Heat can make things move by producing pushing and pulling power. Most engines burn *fuels* such as oil, gasoline, paraffin, or kerosene, to produce heat. The heat then produces the pushes that cause the engine to work.

Look at these pictures. All of these vehicles have engines that burn fuels to produce heat. Say what fuel you think each of these vehicles would use.

SPIRALING SNAKES

You can see the effect of heat quite easily.

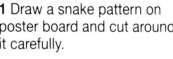

You will need: a piece of poster board; colored pencils; scissors; thread.

1 Draw a snake pattern on poster board and cut around it carefully.

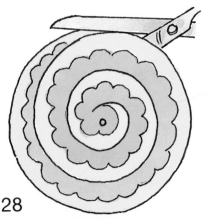

2 Hang it from a thread.

3 Ask an adult to hang your spiral just above the bulb of a lighted table lamp. Or, carefully, hold the spiral above a hot radiator. Watch to see what happens.

PAPER WAND

Now try this experiment.

1 Cut some narrow strips of tissue paper.

2 Fix them to the end of a stick with tape.

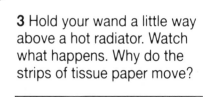

3 Hold your wand a little way above a hot radiator. Watch what happens. Why do the strips of tissue paper move?

NEVER touch hot bulbs or radiators as they can burn.

BOUNCING BEADS

In the last project, the spiral and strips of tissue paper moved because the lamp or radiator heated the air. The hot air rose, pushing the paper.

There is another way you can see how heat makes things move. Ask an adult to put a saucepan of water on a stove. A glass saucepan is best, but any kind will do.

1 Put a handful of glass beads in the water, or a handful of pea-sized pebbles. Do NOT use plastic beads.

2 Rest the lid loosely on the saucepan. Then ask an adult to turn on the heat. Watch what happens.

When the water gets hot you will hear the beads or pebbles bouncing around inside the saucepan. The lid of the saucepan may even be pushed up and down.

Heat from the stove makes the water turn into steam. When the water becomes steam, it gets larger or *expands* to take up more room. The steam pushes on the lid of the saucepan, moving it up and down. The pushing also makes the water bubble and the beads move around.

This pushing by steam is what makes a steam engine work.

The Jumping Tin Lid

Show this experiment to your friends. Ask them to say how it works. You need an empty tea tin with a lid.

1 Put the lid on the tin. Do not push it on too hard.

2 Stand the tin in the bottom of a bowl. Ask an adult to pour hot, NOT boiling, water carefully into the bowl.

Watch the lid of the tin. What happens? Can you explain what you see?

ENERGY FROM THE SUN

Most of the energy on earth comes from the sun. Even the energy we use to run, walk, talk, and swim came originally from the sun.

Light energy from the sun makes plants grow. We eat some of these plants. The meat and other foods we eat such as milk and eggs come from animals that feed on plants which get their energy from the sun. Two kinds of energy come from the sun – light and heat.

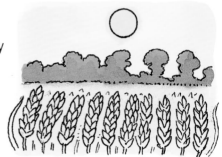

HEAT TRAP

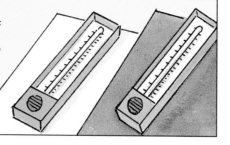

You will need: 2 pieces of paper, both the same size, one black and the other white; 2 thermometers.

1 Lay the *thermometers* side by side in a sunny place. Record the temperature of each one. Cover one thermometer with black paper, and the other with white paper.

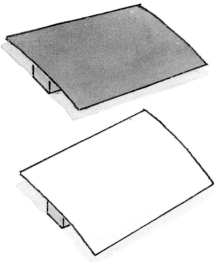

2 After 15 to 30 minutes, remove both pieces of paper. Quickly read the temperature of each thermometer.

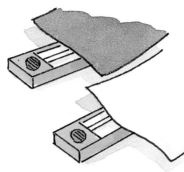

Under which paper was it hottest, the black paper or the white? Which color *absorbs*, or soaks up, more of the sun's heat.

What color clothes would you wear if you wanted to stay cool on a sunny day?

Solar Panels

This is a *solar* panel. It heats water for use in the house below. The panel absorbs some of the sun's energy and heats water running through tiny tubes.

Why are solar panels like this a good idea? Why are they better than, say, oil or gas heaters? In what ways are they not as good?

SOLAR HEATING

1 Find two or three clean plastic bottles, all the same size. Fill them all with the same amount of cold water.

2 Stand the bottles in a row on a sunny windowsill. Cover each with a different colored paper.

3 Leave the bottles for 30 minutes. Then see which bottle has the warmest water in it. Find out with a thermometer or simply by dipping your finger in each one.

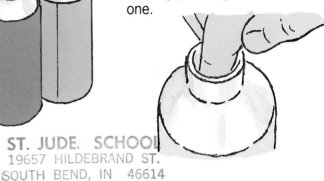

SOLAR POWER

It is possible to concentrate the sun's rays and make them stronger. A solar cooker does this to cook food. Make your own solar cooker and try it out on a hot sunny day.

You will need: metal cooking foil; a cardboard shoe box; poster board; glue or tape; piece of wire from a coat hanger; a hot dog or some other food.

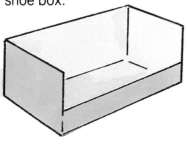

1 Cut the front from a strong shoe box.

2 Cut a disk from the poster board. Cut it in half.

3 Use a rectangle of poster board and the two semicircles to make a trough shape.

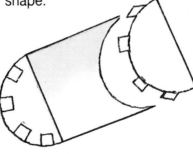

4 Line the inside of the trough with cooking foil.

5 Attach the trough, and your food, to the shoe box with the wire.

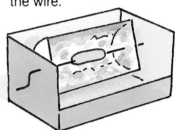

6 Place the cooker in the sun. Keep turning your food so that it cooks evenly.

ALWAYS make sure all food is cooked thoroughly. Ask an adult to test it.

ENERGY FROM FOOD

Plants are the only living things that can use the sun's energy to make their food.

Coal, from which we obtain a great deal of energy, was made by green plants that used the sun's energy millions of years ago. Oil and gas were made by tiny animals called plankton that fed on plants that used the sun's energy millions of years ago. So our bodies, and all vehicles, are worked by energy that once came from the sun.

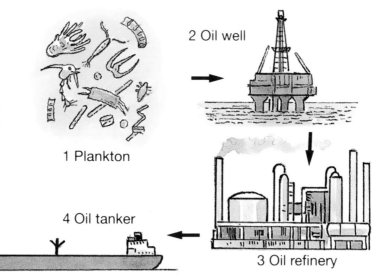

1 Plankton

2 Oil well

3 Oil refinery

4 Oil tanker

FATTY FOODS

Foods that give most energy are *fats* and the different types of cooking oil. Find out which foods contain fat or oil by doing this simple test.

1 Rub a little piece of butter or margarine on a piece of white paper. Hold the paper up to the light.

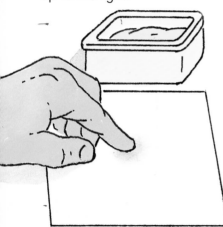

Where you rubbed the fatty food, the paper will let light through. This spot is said to be *translucent*.

2 Test lots of different foods by rubbing little pieces onto paper and make a chart of your results.

Foods and Energy

Many people get a lot of their food energy from sugar and flour. Both sugar and flour are made from plants.

In our bodies, the muscles use these energy foods to keep us alive and able to move.

FLOUR POWER

How can we get the energy out of sugar and flour without eating them?

You will need: a large metal tray; sand; short candle; thick metal foil; flour; sugar; pair of pliers; strong wooden clothes pin.

1 Fill the large metal tray with sand. Stand the candle in the middle.

2 Make a small spoon shape from a strip of thick metal foil.

3 Put a tiny pinch of flour in the spoon and then hold it in a pair of pliers or a strong wooden clothes pin.

4 Ask an adult to light the candle. Carefully hold the spoon with the flour in it over the flame. Can you get heat or light from the flour?

5 Clean the spoon and repeat the experiment using sugar. What happens?

You could try this with other foods which seem floury or sugary.

Labeled!

Make a collection of labels from food packages and cans. The labels will show how much of the different foods – fats, *carbohydrates*, and *proteins* – there are in the packages or cans.

Carbohydrates are floury or sugary foods that give us energy. Proteins such as eggs and meat help us to grow. Some foods contain vitamins and minerals. What do these do?

GLOSSARY

Here are the meanings of some words you might have met for the first time in this book.

ABSORB: to soak or suck up; to take in.

BALANCE: an instrument used for weighing.

CARBOHYDRATES: starches, sugars, and floury substances in food.

CONTRACT: to make or become smaller.

ENERGY: something that animals and machines must have if they are to do work. Food provides energy for animals, including people, while fuels provide energy for machines.

EXPAND: to become larger.

FATS: substances found in food or used in cooking, such as butter, margarine, and lard.

FORCES: the pushes and pulls that can start or stop the movement of an object, change its direction when it is already moving, or change its shape.

FRICTION: the rubbing force that slows down or stops movement.

FUEL: a material that is used to produce heat or power by burning.

GRAVITY: the invisible force that pulls things toward the Earth and gives them weight.

HEAT: warmth, the feeling received from the sun, a fire, or a radiator.

INERTIA: the tendency of something to stay still or keep moving. To make an object start or stop moving you have to overcome its inertia by pushing or pulling it.

KINETIC ENERGY: the energy of motion.

MUSCLE: one of the parts of the body used to produce movement.

PENDULUM: a swinging wire or rod with a weight on the end.

POTENTIAL ENERGY: the amount of energy in an object that can be used.

PROTEIN: a substance in food which is necessary to make the body grow.

PULLEY: a wheel with a groove in the rim for a rope or chain to run over. It is used for lifting heavy things with a small amount of force.

RAMP: a slope joining two levels.

RELAX: to unbend; to become less rigid.

SOLAR: of or from the sun; using the sun's rays.

THERMOMETER: an instrument used for measuring temperature.

TRANSLUCENT: letting light through but not enough to see through.

WEIGHT: the measured heaviness of something; the effect of the force of gravity pulling on something.